Kay's Multiple Choice Questions in Chemical Pat

Kayode (Kay) Adebayo

MB BCh(Ife Nig), MSc & MHRS(Lagos Nig), MCRRA(Chicago)

Published by IISTE

ISBN: 978-1-62265-970-8

ISBN-10: 1622659708

Contents

Respond to each option by indicating whether true (T) or false (F)

Chapter 1   Block 1 questions

1.Causes of hyperkalaemia include:
- A.   Vomiting
- B.   Diarrhoea
- C   Renal failure
- D   Diuretic therapy
- E   Addison's disease

2.Total parenteral nutrition could be complicated by:
- (A)   hyponatraemia
- (B)   hypoglycaemia
- (C)   hyperinsulinaemia
- (D)   hypophosphataemia
- (E)   secondary hyperparathyroidism

3.The tumour lysis syndrome can result in:
- (A)   renal failure
- (B)   hyperkalaemia
- (C)   hypouricaemia
- (D)   hypercalcaemia
- (E)   hyperphosphataemia

4.Hypercholesterolaemias are commonly associated with:
- (A)   thyrotoxicosis
- (B)   obstructive jaundice
- (C)   nephrotic syndrome
- (D)   excessive alcohol intake
- (E)   lipoprotein lipase deficiency

5.Causes of lactic acidosis include:
- (A)   sepsis
- (B)   liver failure
- (C)   renal tubular acidosis
- (D)   glycogen storage disease
- (E)   mitochondrial DNA disease

6.The following are transported in plasma by binding onto albumin:
- (A) Cu^{2+}
- (B) Fe^{3+}
- (C)   free fatty acids
- (D)   tri-iodothyronine (T_3)
- (E)   unconjugated bilirubin

7.An episode of acute extra-vascular haemolysis is associated with:
- (A)   bilirubinuria
- (B)   haemoglobinuria

(C)   hyperurobilinogenuria
(D)   unconjugated hyperbilirubinaemia
(E)   low serum concentrations of haptoglobin

8.A raised plasma alkaline phosphatase activity concentration is helpful for the diagnosis of:
(A)   viral hepatitis
(B)   myelomatosis
(C)   Paget¡¦s disease
(D)   haemolytic jaundice
(E)   pulmonary infarction

9.Laboratory features of water depletion include:
(A)   haemodilution
(B)   hypernatraemia
(C)   low plasma osmolality
(D)   high plasma ratio of creatinine/urea
(E)   high concentration of plasma albumin concentration

10.The following changes related to calcium homeostasis occur in chronic renal failure:
(A)   vitamin D resistance
(B)   hyperphosphataemia
(C)   metastatic calcification
(D)   primary hyperparathyroidism
(E)   secondary hyperparathyroidism

11.The following are features of Cushing's syndrome:
(A)   hypertension
(B)   hypokalaemia
(C)   metabolic acidosis
(D)   impaired glucose tolerance
(E)   non-suppressible serum cortisol level after low dose dexamethasone suppression test

12.The following analytes would rise in concentration with the development of diabetic ketoacidosis:
(A)   blood pH
(B)   serum creatinine
(C)   serum phosphate
(D)   serum potassium
(E)   urine β-hydroxybutyrate

13.Serum prostate specific antigen (PSA) is:
(A)   an enzyme
(B)   an inappropriate tumour marker
(C)   not present in female
(D)   specific to prostatic cancer
(E)   an isoenzyme of prostatic acid phosphatase

14.The following is/are usually seen in relation to ectopic ACTH secretion:
(A)   hyperkalaemia
(B)   metabolic acidosis

(C)   glucose intolerance
(D)   skin hyperpigmentation
(E)   symptoms may occur within days or weeks

15.A low plasma albumin concentration is frequently seen in:
  (A)   acute viral hepatitis
  (B)   nephrotic syndrome
  (C)   severe burns patients
  (D)   carcinoma of the stomach
  (E)   inappropriate secretion of antidiuretic hormone (SIADH)
16.Hyponatraemia is commonly found in patients with:
  (A)   diabetes insipidus
  (B)   sick cell syndrome
  (C)   alcoholic liver disease
  (D)   with congestive cardiac failure
  (E)   urinary protein output of more than 5 g/day
17.In obstructive jaundice:
  (A)   plasma alanine transaminase is decreased
  (B)   plasma activity of alkaline phosphatase rises
  (C)   plasma total protein concentration is increased
  (D)   plasma conjugated bilirubin concentration rises
  (E)   urinary excretion of urobilinogen increases except in total obstruction
18.The following cardiac markers are likely to be elevated in a specimen drawn 72 hours after an uncomplicated myocardial infarction:
  (A)   LDH
  (B)   CK-MB
  (C)   Myoglobin
  (D)   Troponin T
  (E) $CK-MM_3 / CK-MM_1$ ratio

19.Human chorionic gonadotropin:
  (A)   stimulates the corpus luteum
  (B)   has a subunit similar to TSH
  (C)   is secreted by the syncytiotrophoblast
  (D)   is low in patients with choriocarcinoma
  (E)   is structurally similar to luteinising hormone

20.Haemolysis in the blood specimen is an important cause of elevated serum activity concentration of:
  (A)   alkaline phosphatase (ALP)
  (B)   lactate dehydrogenase (LD)
  (C)   alanine transaminase (ALT)
  (D)   total acid phosphatase (ACP)
  (E)   aspartate transaminase (AST)

21.Pathophysiological processes with disturbed hydrogen ion input/output rate can be excluded by:
  (A)   a normal $PaCO_2$ value

(B)   a normal arterial pH value
(C)   a normal arterial bicarbonate value
(D)   normal pH, PaCO$_2$, and arterial bicarbonate values
(E)   normal pH, PaCO$_2$, arterial bicarbonate and anion gap values

22. Serum urate concentration:
(A)   can be elevated in renal failure
(B)   falls in the first trimester of pregnancy
(C)   is always elevated in patients with gout
(D)   is usually higher in males than in females
(E)   is reduced in myeloproliferative disorders

23. Cholestyramine lowers blood cholesterol concentration by:
(A)   feedback inhibition of endogenous synthesis
(B)   hormonal repression of hepatic LDL production
(C)   promoting cholesterol uptake by peripheral tissue
(D)   sequestering bile salts and thus decreasing intestinal absorption
(E)   induction of lecithin-cholesterol acyltransferase (LCAT) enzyme

24. Regarding acute intermittent porphyria:
(A)   Blood lead of the patient is increased
(B)   RBC porphyrin of the patient is increased
(C)   photosensitivity is one of the clinical features
(D)   abdominal pain is a common presenting feature
(E)   there is an elevation of urinary porphobilinogen excretion

25. Regarding congenital hypothyroidism:

(A)   the incidence is about 1 in 30,000 live births
(B)   mental retardation is a complication of delayed treatment
(C)   the condition may be transient and recover spontaneously
(D)   Nigeria has a neonatal screening programme for the condition
(E)   thyroid gland dysgenesis/hypoplasia is commoner than secondary hypothyroidism

26. Regarding the Syndrome of Inappropriate Antidiuretic Hormone secretion:
(A)   the adrenal function is always normal
(B)   urinary osmolality is inappropriately high
(C)   measurement of urinary concentration of sodium is useful
(D)   mutation in the vasopressin receptor has been demonstrated
(E)   measurement of the plasma antidiuretic hormone is clinically helpful for most cases

27. Chronic pancreatitis can account for:
(A)   an increased faecal fat excretion
(B)   abnormal results in an oral xylose absorption test
(C)   a significant increase in serum alpha-amylase activity
(D)   low activity of trypsin (below 20units) to Lundh meal
(E)   low urine specific radioactivity in all 3 stages of the Schilling¡¦s test

28.During the first week of life, neonates born at term have a:
(A)   higher plasma creatinine level than adult
(B)   higher glomerular filtration rate than adult
(C)   lower plasma bicarbonate concentration than adult
(D)   lower renal fractional excretion of sodium than adult
(E)   the ability to achieve a maximum urine concentration of around 600 mmol/kg

29.In phaeochromocytoma:
(A)   the tumour can be found in the urinary bladder
(B)   the tumour is of neural crest origin embryologically
(C)   the tumour may be bilateral involving both adrenal glands
(D)   the majority of tumours secrete noradrenaline more than adrenaline
(E)   it is one of the components of the MEN 2a (multiple endocrine neoplasia) syndrome

30.In hypertension:
(A)   renal artery stenosis results in secondary hyperaldosteronism
(B)   mineralocorticoid excess results in excessive renal loss of potassium
(C)   95% of cases of hypertension are classified as essential hypertension
(D)   Conn¡¦s syndrome is the most common secondary cause for hypertension
(E)   secondary causes need to be excluded in hypertensive patients with a young age-of-onset

31.A urine sample:
(A)   preserved with alkali is suitable for albustix
(B)   can be used for diagnosing light chain disease
(C)   preserved with acid leads to raised levels of urinary free cortisol
(D)   preserved with acid is not suitable for the determination of catecholamines
(E)   has a higher content of large molecular weight proteins in non-selective nephrotic syndrome

32.The following statement(s) is/are true:
(A)   hypoglycaemia is a feature in glycogen storage disease
(B)   high plasma lactate is found in glycogen storage disease type I
(C)   urine reducing substance is present in patients with galactosaemia
(D)   serum free carnitine is increased in patients with organic acidaemia
(E)   large amount of ketone bodies are formed during hypoglycaemic attacks in patients with fatty acids ß-oxidation defects

33.Regarding pre-analytical factors which can influence the Clinical Chemistry results
:(A)   treatment with oestrogens will lead to raised levels of total thyroxine
(B)   treatment with prednisolone yields low levels of cortisol by immunoassay
(C)   blood stored at $4^{o}C$ overnight can lead to increased level of plasma potassium
(D)   blood collected in EDTA gives low laboratory results for alkaline phosphatase
(E)   heparinized blood removes fibrinogen, thereby reducing interference in protein electrophoresis

34.Regarding acute renal failure:
(A)   prostatic hypertrophy may be a cause in some patients
(B)   acute renal failure can be a complication of crush injuries
(C)   in pre-renal azotaemia, the fractional excretion of sodium is usually less than 1%

(D)   the mortality for acute tubular necrosis is less than acute renal failure due to pre-renal causes

(E)   casts and cell debris are often seen in the urinary sediment of patients with pre-renal azotaemia

35. With respect to inborn errors of copper and iron metabolism:
   (A)   patients with hereditary haemochromatosis have low serum iron concentrations
   (B)   in Wilson¡¦s disease, the accumulation of copper in the liver leads to liver failure
   (C)   patients with hereditary haemochromatosis have high serum ferritin concentrations
   (D)   patients with Wilson¡¦s disease, have low serum copper and caeruloplasmin concentrations
   (E)   hereditary haemochromatosis is associated with iron accumulation in the liver, leading to liver failure

36. In cerebrospinal fluid (CSF):
   (A)   a blood stained specimen may indicate a recent subarachnoid haemorrhage
   (B)   the glucose concentration is lower than the reference interval in viral meningitis
   (C)   the glucose concentration should normally be 20% less than the plasma glucose concentration
   (D)   small molecular weight proteins are present in lower concentrations than larger molecular weight proteins
   (E) the total protein concentration can be raised in bacterial meningitis, multiple sclerosis and encephalitis

37. In protein electrophoresis:
   (A)   the serum pattern of a nephrotic patient has a marked α-2 band
   (B)   caeruloplasmin and haptoglobin have mobility in the α-1 region
   (C)   the albumin and transferrin bands are normally increased in a patient with acute phase reaction
   (D)   the deficiency of α-1 antitrypsin is characterized by a marked increase in the respective band
   (E)   the physiological immunoglobulin bands are usually suppressed in the serum of a myeloma patient

38. Regarding prenatal diagnosis:
   (A)   counselling should be given prior to the diagnostic procedure
   (B)   karyotyping can be performed using materials obtained from both chorionic villus sampling and amniocentesis
   (C)   amniocentesis is commonly performed in the first trimester to ensure a prompt provision of diagnostic results
   (D)   the amniotic fluid contains a mixture of fetal and maternal cells which require separation prior to DNA analysis
   (E)   analysis of amniotic fluid using the polymerase chain reaction is generally carried out following a period of culturing

39. Regarding the performance characteristic of a laboratory test:
   (A)   prevalence of the disease has no bearing on the predictive value of a test
   (B)   sensitivity is the proportion of patients with the disease who are identified by the test

(C)   specificity is the proportion of patients without the disease who are identified by the test

(D)   adjustment of the diagnostic cut-off value to improve the sensitivity will usually improve specificity as well

(E)   ROC (receiver operating characteristic) curves are useful in the assessment of two or more tests for a given condition

40.Regarding the water deprivation test:

(A)   it is useful for the investigation of certain genetic disorders of water metabolism

(B)   the patient is bled regularly for the assessment of plasma antidiuretic hormone level.

(C)   the patient¡¦s weight is no longer used as a monitoring parameter because it is superseded by newer tests.

(D)   it can be used to help to localise the anatomical position of the pathologies causing certain disorders of water metabolism.

(E)   the patient is encouraged to drink as much water as possible prior to the start of the test to prevent dehydration during the testing period.

41   The physiologic blood PH in man is maintained at
     a) 35 – 45nmol/L
     b) 25 – 30nmol/L
     c) 135 – 145nmol/L
     d) 125 – 135nmol/L
     e) 35 – 45mol/L

42   Sources of acids in humans include the following except;
     a) metabolism of amino acids
     b) incomplete metabolism of carbohydrate compounds
     c) foods
     d) drugs
     e) none of the above

43   Blood buffers include
     a) HHb/O_2Hb
     b) HCO_3^-/H_2CO_3
     c) Lactic acid/ lactate
     d) Hb/plasma protein
     e) H_2SO_4/SO_4^{2-}

44   Which of the following is wrongly paired
     a) syndrome ---------SG 1.010,hypoelectrolytaemia
     b) Bicarbonate reclamation ----proximal convoluted tubules
     c) Hypernatraemia -------Haemoconcentration
     d) Blood PH 6.1 ----------acidosis
     e) Blood PH 8.5 ----------alkalaemia
     1

45   Biochemical tumour markers include
     a) Oncofetal proteins eg CEA, AFP, HCG
     b) Mucin glycoproteins eg Ca125, Ca15-3,CD30
     c) Enzymes eg PSA, PAP, LDH
     d) Hormones and receptors eg oestrogen receptors, ACTH

e) Cell surface proteins

46   Causes of overflow proteinuria include

a) lysozymuria
b) myoglobinuria
c) microalbuminuria
d) Bence Jones proteinuria
e) Tamm-Horsfall proteinuria

47   Factors predisposing to urinary calculi formation include:

a) cystinuria
b) hypocalciuria
c) hyperoxaluria
d) hypercitraturia
e) renal tubular acidosis

48   Hypercalcaemia is associated with

a) myeloma
b) sarcoidosis
c) hypervitaminosis D
d) secondary hyperparathyroidism
e) excessive absorbable alkali consumption

49.   Application of tumour markers include

a) detection or screening
b) misdiagnosis
c) monitoring and classification
d) staging
e) therapy

50.   Concerning production of tumour markers

a) ACTH from Oat cell Ca of lungs is appropriate
b) ADH from   "   "   "   ' is appropriate
c) Calcitonin from medullary Ca of thyroid is inappropriate
d) Adrenalin from neuroblastoma is inappropriate
e) TSH from hydatidiform mole is appropriate

Chapter 2   Block 2 questions

1. Chemical pathology;

   a) Also known as clinical biochemistry and clinical chemistry
   b) Came into Nigeria with the establishment of the medical faculty of the University of Ibadan.
   c) Came to Nigeria in early 1970's.
   d) Earliest Nigerian in this field are Osafor , Aladenika and Edozien
   e) Deals essentially with result generation in the laboratory

2. Specimens received in the chemical pathology laboratory;
   a) include sweat, urine, stool, sputum and blood
   b) specimens collected with EDTA bottles cannot be used
   c) Labeling of sample bottles correctly and filling of forms are not mandatory.
   d) The skin can be cleaned with 45% alcohol, Hibitane or chlorexidine before venepuncture.
   e) Blood can be collected from the vein, artery or capillary.

3. Pre-analytical variables include;
   a) posture
   b) diurnal variation
   c) icthymal variation
   d) age, sex, race and height
   e) genotype and blood group of the client/patient

4. Concerning water and sodium balance;
   a) homeostatic mechanisms of water and sodium are interlinked
   b) aldosterone secretion is the most important factor affecting body sodium content
   c) ADH secretion is controlled by changes in plasma osmolality and contracted ECF volume
   d) Plasma colloid osmotic pressure depends on sodium concentration
   e) Water depletion may be caused by hypotonic intravenous infusion

5. Potassium metabolism;
   a) hyperkalaemia can be caused by haemolysis, leucocytosis and thrombocytosis.
   b) Hypokalaemia can be caused by Fanconi syndrome, renal tubular acidosis,etc.
   c) Hyperkalemia is usually a feature of Addison's disease.
   d) Hyperkalaemia is usually a feature of prolonged vomiting.
   e) Arrythmias and hypertension can result from hypokalaemia.

6. The kidney;
   a) excretes waste products of metabolism
   b) central in hydrogen ion homeostasis

   c) the functional unit is the glomerulus which is up to one million
   d) filteration is a function of molecular weight of protein, the electrical charge as well as the size of the pore.
   e) The distal convoluted tubule is the site for absorption of most glomerular filtrate- glucose, amino acids, bicarbonate, etc.

7. Functions of the kidney;
   a) production of erythropoietin (EPO)
   b) absorption of phosphate a process inhibited by PTH
   c) absorption of bicarbonate actively in PCT
   d) absorption of calcium through nephrons by stimulation of PTH
   e) endocrine functions …production of rennin, calcitriol and pepsin.

8. The liver;
   a) weighs about 1,300-1,600g in an adult
   b) Its functions include metabolic, synthetic, storage excretory and catabolic.
   c) Potted liver could be ante-mortem or post-mortem
   d) Storage functions include that of glycogen, triglyceride, Iron, Copper and Sodium.
   e) Acinus zoning is based on the classical hexagonal structure while lobular arrangement is based on hepatic microcirculation.

9. A 58 year old woman presented at ISTH with a generalized excoriating rash. This had Been present and progressive for 3years. on examination she was clinically mildly jaundiced without hepatomegally. No urobilinogen in the urine but bilirubin was present. Total Bilirubin was 54µmol/L(<17),ALT was 36U/L(<50) AND ALP was 613U/L.

   a) there is moderately elevated plasma total bilirubin with raised ALP
   b) urinalysis with raised plasma Tbil suggests hepatobiliary disorder
   c) there is haemolysis with prehepatic hyperbilirubinaemia.
   d) the history, examination, and investigation gives a picture of hepatitis.
   e) Differentials include ascending cholangitis and Ca head of pancreas.

10. Concerning porphyrias;
   a) they are acquired metabolic disorders of porphyrin metabolism.
   b) There is an excess haem production
   c) Generally there is an excess production of ALA, PBG and porphyrins
   d) Hepatic porphyries include acute intermittent porphyria, hereditary corproporphyria and porphyria variegate.
   e) Porphrinurias may be due to lead poisoning and liver disease.

11. Which of the following is/are incorrectly paired?
   a) chylomicrons   apolipoproteins B-48,C,A,E

b) IDL   Apolipoproteins B-100,E

c) Lipoprotein (a)   Apolipoproteins B-100, (a), albumin

d) HDL   Apolipoproteins B,C,E

e) VLDL   Apolipoproteins B-100, A,C.

12. Hormones:

a) Have non specific actions

b) Always produced by ductless glands

c) Have independent actions

d) One hormone may have many actions

e) Necessary for reproduction.

13. Steroid hormones:
   a. Interact with extracellular receptors
   b. Interact with intra nuclear receptors
   c. Thyroid hormones are examples
   d. May lead to protein hormone synthesis
   e. May produce a secretory product.

14. Actions of progesterone include:
   a) Promotes ovulation
   b) Production of corpus luteum
   c) Increase male body temperature
   d) Maintenance of early pregnancy
   e) Support breast lactation

15. Theca/stromal cells of the ovary
   a) Produce androgens
   b) Produce estrogens
   c) Produce progesterone
   d) A is true while C is False
   e) None of the above is True

16. The female menstrual cycle;
   a) In follicular phase LH and FSH increases
   b) Regeneration of the endometrium
   c) Regression of the endometrium
   d) There is LH surge

e) The follicules may rupture in this phase

17. Infertility;
   a) Is voluntary failure to conceive
   b) Is valid after 2years of unprotected sex
   c) Is secondary when the cause is hormonal
   d) Is primary when the cause is uterine factors
   e) Is common with long distance truck drivers.

18. Facts about infertility:
   a) Male factors accounts for 40% of cases
   b) Female factors accounts for 50% of cases
   c) Female and male factors accounts for 10%
   d) May be unexplained
   e) It decreases with advancing age.

19. Essential nutrients include;
   a) All amino acids
   b) All water soluble vitamins
   c) Vitamins A & D
   d) Vitamins A & E
   e) Vitamins E & K

20. Requirements for nutrients varies according to:
   a) age
   b) body mass index
   c) reproductive status
   d) Disease state
   e) Therapeutic intervention

21. Difference(s) between Kwashiorkor and Marasmus,
   a) Blood glucose is low in marasmus
   b) Total protein is low in kwashiorkor
   c) Albumin is low in both
   d) Hyponatraemia is present in both of them
   e) A is true while B is false

22. Patients for parenteral nutrients;
   a) Severely burnt patients
   b) Severe dysphagia
   c) persistent vomiting
   d) Short bowel syndrome
   e) Comatose patients

23. Causes of polyuria may include:
   a) Alcoholism

    b) Diabetes mellitus
    c) Hypercalcaemia
    d) Hyperlipidaemia
    e) Diabetes insipidus

24. Urine examination:
    a) Liver function test may include urinalysis
    b) Excess urobilinogen in SCA patients
    c) RBC casts is found in nephrotic patients
    d) Proteinuria may occur in hypertension
    e) Oliguria is common in acute renal failure.

25. Autosomal dominant inheritance:
    a) Every affected individual has one affected parent
    b) Normal offspring may be carriers
    c) All children are affected if both parents are heterozygous
    d) Successive generation may be spared
    e) A is true while D is false.

26. In kinetic enzyme assay, rate of activity:
    a) Maximum in first order kinetic phase
    b) Maximum substrate depletion phase
    c) Clinical assays usually follow first order kinetics
    d) Absorbance is best measured during the linear phase
    e) A is true while C is false

27. In the third trimester of a normal pregnancy;
    a) ALP is elevated
    b) ALP is reduced
    c) AST is elevated
    d) Gamma GT is elevated
    e) Albumin is reduced

28. Ehis is a teenager with fractured femur and hepatitis, therefore;
    a) AST is elevated
    b) Lactate dehydrogenase is normal
    c) Gamma GT is normal
    d) Creatinine kinase is elevated
    e) Alkaline phosphatase is reduced

29. Causes of both ALT and AST elevation are the following EXCEPT;
    a) Congestive cardiac failure
    b) Cirrhosis of the liver
    c) Biliary obstruction
    d) Skeletal muscle injury

e) Acute myocardial infarction

30. Tissues containing significant amount of Gamma glutamyl transferase;
    a) Kidney
    b) Liver
    c) Pancreas
    d) Intestine
    e) All of the above

31. 5' Nucleotidase enzyme is;
    a) Phosphatase enzyme
    b) Elevated in acute liver disease
    c) Elevated in pregnancy
    d) Involved in energy transfer
    e) A is True while C is False.

32. Concerning acid-base balance;
    a) Strong acids have pK values less than 3, while strong bases have pK >9
    b) The body produces 15-20nmol of H^+ each day
    c) ECF acidity ranges from 35-45nmol/L
    d) HCO_3^-/H_2CO_3 buffer is effective because of its pK of 6.1
    e) Buffer base is 46-52mmol/L $\pm$ 3.0mmol/L

33. Some causes of non-respiratory acidosis include;
    a) Ketoacidosis- diabetic/alcoholic
    b) Salicylate poisoning
    c) Renal tubular acidosis types 1 & 4
    d) Diarrhoea and pancreatic fistula
    e) Intravenous feeding with excess cationic amino acids

34. Factors affecting protein requirement and intake include;
    a) Body weight or size
    b) Prosperity and local custom
    c) Age
    d) Sex
    e) Metabolic demand eg pregnancy and lactation

35. Concerning protein electrophoresis;
    a) Separates proteins based on molecular size and electrical charge
    b) Components include electricity source, buffer, template, electrode, and reading or detection mechanism.
    c) Stains are Ponceau S, Sudan red and oil black
    d) Main electrophoretogram bands are albumin, α_1, α_2, β, β_1 and γ
    e) Paraprotein patterns can occur in either β, or γ

36. Plasma proteins:
    a) Functions in tissue nutrition, hormones, receptors, biocatalysts and antibodies
    b) Have similar physicochemical characteristics and physiological roles.
    c) Some are synthesized in the liver controlled by specific regulatory mechanism
    d) Synthesis of acute phase proteins is controlled by humoral factors released from macrophages
    e) Breakdown occurs at ALL capillary endothelial cells through pinocytosis across cells releasing amino acids to the tissues.

37. Regarding calcium homeostasis;
    a) About 99% of body calcium resides in bones
    b) Acidosis causes an increased free ionized calcium
    c) Daily calcium requirement depends on age, growth spurt, pregnancy and lactation.
    d) Plasma calcium is lower in the erect than the supine position.
    e) Calcium activates muscle fibre in smooth muscle.

38. Water and electrolyte balance involves;
    a) Atrial natriuretc peptide (ANP) released from cardiac atria in response to stretch and from ventricles in heart failure
    b) Brain natriuretic peptide(BNP) released from the myocardium and ventricles
    c) C-type natriuretic peptide synthesized in the vascular endothelial cells and brain
    d) Urodilatin is a peptide similar to ATP-ase
    e) Na^+K^+-ATPase inhibitory substance inhibits sodium pump in renal tubules

39. An ideal tumour marker must not have the following properties;
    a) Easy and inexpensive to measure in readily available body fluids
    b) Specific to the tumour in question and commonly associated with it
    c) Have an inverse stoichiometric relationship with the tumour mass
    d) Micrometastasis must be detectable without clinical symptoms
    e) Never detectable in individuals or patients without the tumour.

40. Of tumour markers;
    a) 5-hydroxytryptamine from carcinoid syndrome is inappropriate
    b) Gonadotrophin from hepatocellular Ca is inappropriate
    c) Biochemical classification include mucin glycoproteins, enzymes, oncofetal proteins, hormones and receptors and cell surface proteins
    d) Genetic markers include BCRA1 and 2, RB1 and C erb-2 gene on chr.17
    e) Viral markers include EBV, HIV, HBV and Mokola virus.

41. Quality assurance in chemical pathology laboratory;
    a) This means total quality control of the entire procedure
    b) Accuracy is the test of how specific and sensitive a test is
    c) Specificity is the ability to detect that an analyte is absent when absent
    d) Sensitivity is the ability to detect an analyte when present

e) A reproducible test is good but a repeatable one is a better assessment

42. Errors in scientific measurement;
   a) Include avoidable or systematic errors
   b) Random errors include labeling, containers, temperature and reagent errors
   c) The smaller the SD, the greater the imprecision
   d) In normally distributed measurand, the mean,median and mode occupy same spot
   e) Can be controlled by use of standards, control specimens, and control charts.

43. Protein Calorie/energy malnutrition;
   a) Commoner in females than males
   b) Affects neonates, infants and children (months-5years)
   c) Wellcome classification is based on height, weight in relation to age ± oedema
   d) Aetiology is inadequate protein and calorie intake, aflatoxin ingestion and free radical
   e) Complications include hypocalcaemia, noma,anaemia and hypothermia

44. The fetal-placental unit can be monitored by;
   a) Measuring the hCG and human placental lactogen
   b) AFP level in maternal urine
   c) Maternal bilirubin level in cases of Rh and ABO incompatibility
   d) Lecithin:sphingomyelin ratio determination in amniotic fluid
   e) Amniocenthesis should be done at 12-16week for this assessment

45. Concerning fetal monitoring;
   a) AFP, hCG, Oestriol and Inhibin level are needed to r/o Down's syndrome
   b) L:S ratio of 3 at 34 weeks indicate the need for steroid therapy before delivery
   c) Enzymatic abnormalities can be determined by cell cultures and chionic villi biopsy
   d) High AFP may indicate neural tube defect, exomphalos or multiple pregnancy
   e) LH, prolactin, oestrogen, hCG and FSH continue to rise throughout pregnancy

46. Concerning the hypothalamo-pitutary-gonadal axis;
   a) Hypogonadotrophic hypogonadism may result from malnutrition
   b) Androgens include LH, androstenedione & DHEA
   c) Testosterone from Leydig cells is stimulated by LH
   d) Oestrogen release is cyclinal and highest at menopause
   e) Testosterone and oestrogens circulate bound to albumin

47. Gynaecomastia
   a) Common in African females
   b) May be physiologic in neonates due to maternal oestrogen
   c) At puberty is due to increased oestrogen in relation to androgens
   d) Can be associated with liver disease and hepatic congestion

e) Is a manifestation of virilism

48. Laboratory diagnostic testing performed close to the patient;
    a) Also called value added testing and distributed testing
    b) Can be as out-patient, in-patient and mobile transport and consumer self test.
    c) Has a longer turn around time (TAT)
    d) No test requisition
    e) Instrument is validated, accuracy checked and quality control performed

49. The cerebrospinal fluid;
    a) Secreted by the choroid plexuses and finite capillaries around the cerebral vessels and along the walls of the ventricles of the brain.
    b) Bathes, buoys and feed the brain and the spinal cord
    c) High CSF pressure may be due to CCF, SOL, Cerebral edema, straining
    d) Blood brain barrier index of 200 indicates moderate impairment
    e) CSF lactate level is useful in differentiating between TB and viral meningitis
50. In the elderly,
    a) Diseases include DM, Hypertension, thyroid disease and Paget's disease
    b) Geriatric reference range is easily established like in all age groups
    c) Plasma urea, Creatinine and Urate increases because of aging.
    d) Serum protein increases because of care by children, family and friends
    e) Alkaline phosphatase level increases

Chapter 3   Block 3 questions

(1)   In Acute Liver Disease:

a.   AST/ALT enzymes are markedly elevated

b.   ALP/r-GT are low

c.   Urinalysis is not useful d.Albumin is normal

e. 5^1 Nucleotides is normal

(2)   In Liver Function Test: match the following and mark 'True' or 'False'

a.   ALT/AST $\longrightarrow$ acute hepatitis

b.   Gamma GT $\longrightarrow$ alcohol liver disease

c.   Alkaline phosphatase $\longrightarrow$ obstructive jaundice

d. 5^1 nucleotides $\longrightarrow$ acute hepatitis

e.   ALT and low Albumin $\longrightarrow$ chronic liver disease

(3)   Regan or Nagao Isoenzymes of ALP:

a.   Useful in breast cancer diagnosis

b.   Metastatic carcinoma

c.   Pleural surfaces metastasis

d.   Colonic cancer diagnosis

e.   Brain tumour diagnosis

(4)   In AMI, these enzymes are not elevated:

a.   Lactate dehydrogenate (LD)

b.   Creatinine kinase (CK)

c.   AST

d.   ALP

e.   Gamma GT

(5)   In lactate dehydrogenate Isoenzyme electrophoresis:

a.   LD1 is fastest while LD5 is the slowest

b.   LD5 is the fastest while LD1 is the slowest

c.   LD3 is faster than LD4

d.   LD4 is faster than LD3

e.   None of the above.

(6)   Hormones involved in the menstrual cycle are:

a.   GnRH

b.   LH/FSH

c.   Estrogen

d.   Progesterone

e.   All of the above

(7)   Progesterone:

a.   Synthesized by corpus luteum only

b.   Maintains pregnancy

c.   Increased vascularization of endometrium

d.   Promote ductal breast tissue differentiation

e.   Increases basal body temperature

(8)   During Pregnancy:

a.   Increased plasma hCG

b.   Increased plasma progesterone

c.   Reduced plasma oestrogens

d.   Increased plasma LH/FSH

e.   'A' is true while 'C' is false

(9)   Symptoms of hypothyroidism include all except:

a.   Slow thoughts

b.   Dry hot skin

c.   Constipation

d. Hypocholesterolaemia

e. Heavy menstrual flow.

(10)   Autosomal dominant inheritance pattern: 'T' or 'F'

a. Every affected individual has at least one affected parent

b. Successive generations are affected

c. Normal offsprings are not carriers

d. All of the above are True

e. None of the above is True

(11)   X-linked recessive inheritance pattern:

a. An abnormal X-chromosome becomes latent when combined with X-chromosome

b. Females are typically affected

c. Half of an affected mother's sons are affected

d. A carrier father and normal mother have all their sons affected

e. 'A' is True while 'D' is False

(12)   In serum protein electrophoresis:

a. Serum samples are preferred

b. Plasma samples are preferred

c. Albumin is anodal

d. 5 distinct bands may be seen

e. All of the above.

13. Cholestyramine lowers blood cholesterol concentration by:
    (A)   feedback inhibition of endogenous synthesis
    (B)   hormonal repression of hepatic LDL production
    (C)   promoting cholesterol uptake by peripheral tissue
    (D)   sequestering bile salts and thus decreasing intestinal absorption
    (E)   induction of lecithin-cholesterol acyltransferase (LCAT) enzyme

14. Regarding acute intermittent porphyria:
    (A)   Blood lead of the patient is increased
    (B)   RBC porphyrin of the patient is increased
    (C)   photosensitivity is one of the clinical features
    (D)   abdominal pain is a common presenting feature

(E)   there is an elevation of urinary porphobilinogen excretion

15. Regarding congenital hypothyroidism:

(A)   the incidence is about 1 in 30,000 live births
(B)   mental retardation is a complication of delayed treatment
(C)   the condition may be transient and recover spontaneously
(D)   Nigeria has a neonatal screening programme for the condition
(E)   thyroid gland dysgenesis/hypoplasia is commoner than secondary hypothyroidism

16. Regarding the Syndrome of Inappropriate Antidiuretic Hormone secretion:
(A)   the adrenal function is always normal
(B)   urinary osmolality is inappropriately high
(C)   measurement of urinary concentration of sodium is useful
(D)   mutation in the vasopressin receptor has been demonstrated
(E)   measurement of the plasma antidiuretic hormone is clinically helpful for most cases

17. Chronic pancreatitis can account for:
(A)   an increased faecal fat excretion
(B)   abnormal results in an oral xylose absorption test
(C)   a significant increase in serum alpha-amylase activity
(D)   low activity of trypsin (below 20units) to Lundh meal
(E)   low urine specific radioactivity in all 3 stages of the Schilling¡s test

18. During the first week of life, neonates born at term have a:
(A)   higher plasma creatinine level than adult
(B)   higher glomerular filtration rate than adult
(C)   lower plasma bicarbonate concentration than adult
(D) lower renal fractional excretion of sodium than adult
(E) the ability to achieve a maximum urine concentration of around 600 mmol/kg

19. In phaeochromocytoma:
(A)   the tumour can be found in the urinary bladder
(B)   the tumour is of neural crest origin embryologically
(C)   the tumour may be bilateral involving both adrenal glands
(D)   the majority of tumours secrete noradrenaline more than adrenaline
(E)   it is one of the components of the MEN 2a (multiple endocrine neoplasia) syndrome

20. In hypertension:
(A)   renal artery stenosis results in secondary hyperaldosteronism
(B)   mineralocorticoid excess results in excessive renal loss of potassium
(C)   95% of cases of hypertension are classified as essential hypertension
(D)   Conn¡s syndrome is the most common secondary cause for hypertension
(E)   secondary causes need to be excluded in hypertensive patients with a young age-of-onset

21. A urine sample:
(A)   preserved with alkali is suitable for albustix
(B)   can be used for diagnosing light chain disease

(C)   preserved with acid leads to raised levels of urinary free cortisol

(D)   preserved with acid is not suitable for the determination of catecholamines

(E)   has a higher content of large molecular weight proteins in non-selective nephrotic syndrome

22. The following statement(s) is/are not true:

(A)   hypoglycaemia is a feature in glycogen storage disease

(B)   high plasma lactate is found in glycogen storage disease type I

(C)   urine reducing substance is present in patients with galactosaemia

(D)   serum free carnitine is increased in patients with organic acidaemia

(E)   large amount of ketone bodies are formed during hypoglycaemic attacks in patients with fatty acids ß-oxidation defects

23. Regarding pre-analytical factors which can influence the Clinical Chemistry results:

(A)   treatment with oestrogens will lead to raised levels of total thyroxine

(B)   treatment with prednisolone yields low levels of cortisol by immunoassay

(C)   blood stored at $4^{o}C$ overnight can lead to increased level of plasma potassium

(D)   blood collected in EDTA gives low laboratory results for alkaline phosphatase

(E)   heparinized blood removes fibrinogen, thereby reducing interference in protein electrophoresis

24. Regarding acute renal failure:

(A)   prostatic hypertrophy may be a cause in some patients

(B)   acute renal failure can be a complication of crush injuries

(C)   in pre-renal azotaemia, the fractional excretion of sodium is usually less than 1%

(D)   the mortality for acute tubular necrosis is less than acute renal failure due to pre-renal causes

(E)   casts and cell debris are often seen in the urinary sediment of patients with pre-renal azotaemia

25. With respect to inborn errors of copper and iron metabolism:

(A)   patients with hereditary haemochromatosis have low serum iron concentrations

(B)   in Wilson¡¦s disease, the accumulation of copper in the liver leads to liver failure

(C)   patients with hereditary haemochromatosis have high serum ferritin concentrations

(D)   patients with Wilson¡¦s disease, have low serum copper and caeruloplasmin concentrations

(E)   hereditary haemochromatosis is associated with iron accumulation in the liver, leading to liver failure

26. In cerebrospinal fluid (CSF):

(A)   a blood stained specimen may indicate a recent subarachnoid haemorrhage

(B)   the glucose concentration is lower than the reference interval in viral meningitis

(C)   the glucose concentration should normally be 20% less than the plasma glucose concentration

(D)   small molecular weight proteins are present in lower concentrations than larger molecular weight proteins

(E) the total protein concentration can be raised in bacterial meningitis, multiple sclerosis and encephalitis

27. In protein electrophoresis:

(A)   the serum pattern of a nephrotic patient has a marked α-2 band
(B)   caeruloplasmin and haptoglobin have mobility in the α-1 region
(C)   the albumin and transferrin bands are normally increased in a patient with acute phase reaction
(D)   the deficiency of α-1 antitrypsin is characterized by a marked increase in the respective band
(E)   the physiological immunoglobulin bands are usually suppressed in the serum of a myeloma patient

28. Regarding prenatal diagnosis:
(A)   counselling should be given prior to the diagnostic procedure
(B)   karyotyping can be performed using materials obtained from both chorionic villus sampling and amniocentesis
(C)   amniocentesis is commonly performed in the first trimester to ensure a prompt provision of diagnostic results
(D)   the amniotic fluid contains a mixture of fetal and maternal cells which require separation prior to DNA analysis
(E)   analysis of amniotic fluid using the polymerase chain reaction is generally carried out following a period of culturing

29. Regarding the performance characteristic of a laboratory test:
(A)   prevalence of the disease has no bearing on the predictive value of a test
(B)   sensitivity is the proportion of patients with the disease who are identified by the test
(C)   specificity is the proportion of patients without the disease who are identified by the test
(D)   adjustment of the diagnostic cut-off value to improve the sensitivity will usually improve specificity as well
(E)   ROC (receiver operating characteristic) curves are useful in the assessment of two or more tests for a given condition

30. Regarding the water deprivation test:
(A)   it is useful for the investigation of certain genetic disorders of water metabolism
(B)   the patient is bled regularly for the assessment of plasma antidiuretic hormone level.
(C)   the patient¡s weight is no longer used as a monitoring parameter because it is superseded by newer tests.
(D)   it can be used to help to localise the anatomical position of the pathologies causing certain disorders of water metabolism
(E)   the patient is encouraged to drink as much water as possible prior to the start of the test to prevent dehydration during the test period.

31 The physiologic blood acidity in man is maintained at
          A)35 – 45nmol/L
          B)25 – 30nmol/L
          C)7.51 – 8.51pH
          D)125 – 135nmol/L
          E)7.35 –7.45pH

32   Sources of acids in humans include the following except;

A)metabolism of amino acids
B)incomplete metabolism of carbohydrate compounds
C)foods
D)drugs
E)none of the above

33 Blood buffers include
A)HHb/O_2Hb
B)HCO_3^-/H_2CO_3
C)Lactic acid/ lactate
D)Hb/plasma protein
E)H_2SO_4/SO_4^{2-}

34 Which of the following is not wrongly paired
A)Urea----30mmol/L, Creatinine...10mg/dl-......azotaemia
B)Bicarbonate reclamation ----proximal convoluted tubules
C)Hypernatraemia -------Haemoconcentration
D)Blood PH 6.1 ----------acidosis
E)Blood PH 8.5 ----------alkalaemia

35 Biochemical tumour markers include
A)Oncofetal proteins eg CEA, AFP, HCG
B)Mucin glycoproteins eg Ca125, Ca15-3,CD30
C)Enzymes eg PSA, PAP, LDH
D)Hormones and receptors eg oestrogen receptors, ACTH
E)Cell surface proteins

36 Causes of overflow proteinuria include

A)lysozymuria
B)myoglobinuria
C)microalbuminuria
D)Bence Jones proteinuria
E)Tamm-Horsfall proteinuria

37 Factors predisposing to urinary calculi formation include:
A)cystinuria
B)hypocalciuria
C)hyperoxaluria
D)hypercitraturia
E)renal tubular acidosis

38 Hypercalcaemia is associated with
a)myeloma
b)sarcoidosis
c)hypervitaminosis D
d)secondary hyperparathyroidism
e)excessive absorbable alkali consumption

39. Application of tumour markers include
a)detection or screening
b)misdiagnosis
c)monitoring and classification
d)staging
e)therapy

40. Concerning production of tumour markers

a)ACTH from Oat cell Ca of lungs is inappropriate
b)ADH from   "   "   "   ' is inappropriate
c)Calcitonin from medullary Ca of thyroid is appropriate
d)Adrenalin from neuroblastoma is appropriate
e)TSH from hydatidiform mole is inappropriate

41. Concerning calcium and phosphate metabolism;
   a) Hyperphosphataemia can result from severe burns
   b) Fanconi's syndrome is a cause of hyperphosphataemia
   c) Pseudohypoparathyroidism is a known cause of hypercalcemia.
   d) Tumour lysis syndrome causes hypercalcemia
   e) Parathyroid hormone(PTH) is the principal hormone involved in calcium homeostasis and acts directly on bone and intestine.

42. Concerning water homeostasis;
   a. Antidiuretic hormone(ADH) release through changes in effective circulating volume is a more sensitive and efficient system than ADH release through changes in osmolality.
   b. Syndrome of inappropriate ADH secretion(SIADH) can be diagnosed by the water deprivation test.
   c. Diabetes mellitus is a cause of reduced urine osmolality.
   d. Hypernatremia is a common feature of diabetes insipidus.
   e. Prolonged lithium therapy is a potential cause of nephrogenic diabetes insipidus.

43. Identify the correctly matched lipoprotein and its component apoprotein fraction(s)
   a) Lp(a)   : apoproteins B-100, apoprotein(a)

   b) IDL   : apoproteins B, E

   c) HDL   : apoproteins B, C, E

   d) VLDL : apoproteins A, C, E

   e) LDL   : apoprotein   B-100

44. Concerning gastrointestinal tract(GIT) function;
   a. Xylose absorption test will produce an abnormal response in intestinal causes of malabsorption as against pancreatic causes.
   b. Plasma amylase estimation is a better index of assessment of acute pancreatitis than lipase due to its higher specificity.

   c. Insulin induced hypoglycaemic test is used for investigation of pancreatic function.
   d. Coeliac disease is characterized by anti-gliadin antibodies.

    e. Zollinger-Ellison syndrome is characterized by achlorhydria.

45. In the assessment of foeto-placental integrity in pregnancy;
    a. α-feto protein evaluation can be used to detect neural tube defects and exomphalos
    b. In Down's syndrome the Triple test is characterized by high hCG, low α-fetoprotein and low unconjugated estriol.
    c. Lecthin/sphingomyelin ratio of > 2.0 is indicative of fetal lung maturity.
    d. Neural tube defects are characterized by elevated amniotic fluid acetylcholinesterase.
    e. Multiple pregnancy is associated with elevated hCG levels.

46. All the following are true of Conn's syndrome(I° hyperaldosteronism) except
    a. Hyperkalemia is a common finding
    b. The plasma rennin activity(PRA) is increased
    c) Metabolic acidosis is the common acid-base disturbance

    d) Peripheral oedema is usually present

    f) Nephrotic syndrome is a common cause.

47. In the diagnosis of Cushing's syndrome;

    a) 24hr urine-free cortisol can be used as a screening test.

    b) Definitive diagnosis is made with overnight dexamethasone suppression test.

    c. Cushing's disease will show suppression with high dose dexamethasone suppression test.
    d. Cushing's syndrome due to adrenal tumours is characterized by elevated ACTH levels.
    e. Cushing's syndrome due to adrenal tumours will show suppression following high dose dexamethasone test.

48. Which of these statements is(are) true of lipoproteins ?

    a) The predominant component fraction in VLDL is triglycerides
    b) LDL migrates in the pre-β region on electrophoresis
    c) Each lipoprotein molecule comprises a core of triglycerides and phospholipids.
    d) Apoproteins are useful in the removal of breakdown products of lipoprotein hydrolysis from the circulation.
    e) Apoprotein C-II is the main co-factor for lecithin cholesterol acyl transferase (LCAT).

49. Concerning Addison's disease;
    a. It can be associated with pernicious anaemia
    b. Hypokalemia is a common feature.

    c. Plasma aldosterone levels are normal

    d. In the co-syntropin test, there is and appreciable elevation of cortisol levels from the baseline at 30 and 60minutes.

    e. Hypochloremic metabolic alkalosis is a common feature.

50. Regarding immunoglobins;

    a. IgA has the largest molecular weight

    b. The light chain is responsible for binding effector cells like macrophages

    c. Colostrum is richer in monomeric IgA than dimeric IgA

    d. Hay fever will cause predominatly an elevation in monomeric IgE

    e. Hay fever will cause predominatly an elevation in dimeric IgE

Chapter 4   Block 4 questions

1. Concerning chemical pathology laboratory;

 a) no eating

b) no smoking

c) use of laboratory coats is not a rule

d) bare hands may be used during procedures

e) tap water is the water of choice for reagent preparation.

2. The following specimens can be received in the chemical pathology laboratory;

a)   urine

b)   saliva

c)   faeces

d)   pleural effusion

e)   ascitic tap

3. Preanalytical variables include all EXCEPT;

a)   age

b)   sex

c)   exercise

d)   posture

e)   circadian variation

4. Reference ranges;

a) Sodium is 135-145mmol/L

b) Bilirubin is 1.7-17µmol/L

c) Fasting glucose is 3.5-6.0mmol/L

d) Bicarbonate of 25mmol/L is abnormal

e) Physiologic pH is 7.35 – 7.45

5. Metabolic acidosis may be due to;

a) vomiting

b) retention of CO_2 / COAD

c) renal failure

d) methanol intoxication

e) diarrhea

6.Causes of hypokalaemia include:
   a.   Vomiting
   b   Diarrhoea
   c   Renal failure
   d   Diuretic therapy
   e   Addison's disease

7. 24hours urine specimen may be used in the investigation of;

a) Nephrotic syndrome

b) Pheochromocytoma

c) Acute renal failure

d) Heavy metal poisoning

e) Type II Diabetes Mellitus

8. Reducing sugar may NOT be found in the urine of patients with;

a) Renal failure

b) Thyrotoxicosis

c) Cushing syndrome

d) Galactosaemia

e) Type 2 Diabetes mellitus

9. Factors that could affect enzymatic actions include;

a) Extreme of temperatures

b). Presence or absence of inhibitors

c) pH

d) Enzyme concentration

e) Substrate concentration

10. In acute myocardial infarction :

a) $\underline{LD1}$ ratio < 1 (LD = Lactate Dehydrogenase)

 LD2

b) $\underline{LD1}$ ratio > 1

 LD2

c) CK-MB fraction is elevated within 2hours of AMI

d) LD6 Iso-enzyme may be seen in severe cases

e) All of the above are true.

11. In protein energy malnutrition

a) Serum Zinc is high

b) Serum glucose may be normal in marasmus

c) Total plasma protein is very low in marasmus

d) Plasma albumin is a good assessment tool

e) Pre-albumin is normal

12. Patients who may require nutritional support parenterally include;

a) Unconscious patients

b) Patients with reduced small intestine

c) Patients with persistent and severe dysphagia

d) Patients with Ca stomach

e) Severe burns.

13. Which of these is(are) true of primary hyperaldosteronism?

a) It is characterized by increased urinary potassium excretion

b) It is characterized by increased urinary sodium excretion

c) It is characterized by hypokalemic metabolic alkalosis

d) The plasma renin: aldosterone ratio is increased

e) The plasma aldosterone: rennin ratio is increased

14. The following are causes of hypokalaemia except;

a) prolonged infusion of 50% dextrose

b) prolonged use of salbutamol (β-agonist)

c) Intestinal fistulae

d) Verner-Morrison(WDHA) syndrome

e) Congenital pyloric stenosis

Utilize the history and data below to answer question 15-18

A disoriented patient who presented at the Emergency unit of ISTH was found to have the following parameters;

Na -------136mmol/L		Creatinine ----1.2mg/dl
K -------- 4mmol/L		Ca ------------ 10mg/dl
Urea ---- 24mg/dl		Glucose ------- 90mg/dl
HCO$_3$ ---- 14mmol/L		Chloride ------ 90mmol/L

15. Calculate the plasma osmolarity

a) 284mmol/L

b) 289mmol/L

c) 192mmol/L

d) 254mmol/L

e) None of the above

16. If this patient's measured plasma osmolarity is 320mmol/L, the osmolar gap (if any) is;

a) 30

b) 31

c) 36

d) 66

e) There is no osmolar gap

17. Calculate the patient's Anion gap;

a) 36mmol/L

b) 23mmol/L

c) 31mmol/L

d) 15mmol/L

e) 32mmol/L

18. Which of these can account for the patterns of **both** the anion gap and osmolar gap(if any) simultaneously noted in this patient;

a) Methanol poisoning

b) Salicylate poisoning

c) Ethylene poisoning

d) Uremia

e) Diabetic ketoacidosis

19. Concerning lipoprotein metabolism;

a) The predominant fraction in low density lipoprotein(LDL) is phospholipids

b) The turbidity in a lipaemic sample after a meal is caused by VLDL

c) Lipoprotein(a) has similar lipid concentration to LDL but higher protein content

d) The liver is the main source of endogenous lipids

e) The ligand for the LDL receptor is apolipoprotein B-48

20. In fredrickson's type V hyperlipidaemia, the elevated lipoprotein class(es) include(s);

a) VLDL

b) LDL

c) Chylomicrons

d) HDL

e) IDL

21. Which of these are correctly matched;

a) Chylomicrons : Apolipoproteins B-48,C,A,E

b) IDL   :Apolipoproteins   B-100,E

c) Lipoprotein(a) :Apolipoproteins B-100,(a), albumin

d) HDL   :Apolipoproteins B,C,E

e) VLDL   :Apolipoproteins B-100,A.C

22. Concerning calcium metabolism;

a) Nephrotic syndrome can alter total plasma calcium levels

b) Poture can affect free ionized plasma calcium level

c) Posture can affect total plasma calcium level

d) Liver cirrhosis is most likely to cause an increase in total plasma calcium level

e) Vitamin D_3 intake will cause an increase in faecal phosphate and calcium loss

23. Concerning hyperlipidaemias and other lipid diorders;

a) Familial hypercholesterolaemia is caused by a mutation involving the LDL receptor

b) Hyperthroidism is a very common cause of secondary hyperlipidaemia

c) Hyperthroidism is a very common cause of secondary hyperlipidaemia

d) Tangier's disease is associated with high HDL concentration

e) Chronic alcoholism is a cause of hyperlipidaemia

24. The following are causes of hyperphosphataemia except;

a) Vitamin D toxicity

b) Uremia

c) Granulomatous diseases e.g tuberclosis, sarcoidosis

d) Pseudohypoparathroididsm

e) Tertiary hyperparathyroidism

25. Cushing's syndrome is characterized by all of the following except;

a) Absence of the usual circadian rhythm of cortisol release.

b) Increased urinary loss of potassium

c) Psychiatric disturbances

d) Hyperpigmentation is a strong pointer of adrenal tumour as a likely cause of cushings's syndrome

e) Easy bruisability

26. Which of the following is(are) true of secondary hyperaldosteronism

a) It is characterized by elevated aldosterone levels and high rennin levels

b) it is characterized by elevated aldosterone levels but with low rennin levels

c) It can be caused by pregnancy

d) A low urinary sodium concentration is a feature

e) peripheral oedema is usually absent.

27. Which of these statements is (are) true?

a) Lack of suppression of plasma cortisol levels following low-dose dexamethasone suppression test is diagnostic of Cushing's syndrome

b) Lack of suppression of plasma cortisol levels following high-dose dexamethasone suppression test is diagnostic of Cushing's syndrome

c) A low 24 hour urine-free cortisol is suggestive of Cushing's syndrome

d) Suppression of plasma cortisol levels following an overnight dexamethasone suppression test suggests Cushing's syndrome

28. Concerning congenital adrenal hyperplasia(CAH);

a) hypertension is a common feature of 21α-hydroxylase deficiency

b) hypertension is a common feature of 11β-hydroxylase deficiency

c) Ambiguous genitalia is a feature of males with 17α- hydroxylase deficiency

d) 21α-hydroxylase deficiency is characterized by elevated 17-hydroxyprogesterone levels

e) It is associated with low ACTH levels commonly.

29. The following are tests utilized in the diagnosis of Addison's disease;

a) insulin-induced hypoglycaemic test

b) overnight dexamethasone suppression test

c) Low dose dexamethasone suppression test

d)Tetracosactrin test

e) all of the above.

30. An elderly man who presented to ISTH with symptoms suggestive of osteoarthritis was found on biochemical assessment to have a plasma K level of 5.8mmol/L. which of these is(are)likely cause(s) of this electrolyte pattern this patient?

a) Delay separation of plasma from the whole blood specimen

b) Acute renal failure

c) Spironolactone therapy

d) Digoxin therapy

e) β-adrenergic agonist therapy

31. Which of these pattern is(are) true of the causes of Cushing's syndrome as stated below;

Adrenal tumour	cushing's disease	ectopic ACTH
a) Plasma cortisol		
Elevated	Elevated	elevated
b) 24hr urinary free cortisol		
High	Low	High
c) After low dose dexamethasone test		
No suppression	No suppression	Supressed
d) After high dose dexamethasone test		
No suppression	No suppression	suppressed
e) plasma ACTH		
Low	High	Low

32. Clinical features of Addison's disease include all vexcept;

a) Impaired glucose telorance

b) Truncal obesity

c) Hsypotension

d) Mucosal pigmentation

e) A and D are correct.

33. Which of the following is correct?

a) The body produces 15 – 20mol of H^+ per day

b) H^+ in ECF ranges from 35-45nmol/L

c) H^+ > 45nmol/L can cause neuromuscular irritability, tetany and possible loss of consciousness

d) H^+ <45nmol/L edema and coma

e) About 13-20,000mmol of CO_2 is produced daily

34. The ECF buffers include the following;

a) bicarbonate/carbonic acid

b) hemoglobin

c)plasma proteins

d) ascorbic acid/ascobate

e) erythrocytes and plasma phosphate

35. Concerning renal tubular acidosis(RTA)

a) type I affects the distal tubule and H^+ can not be secreted for excretion

b) type II is due to defect of in proximal tubule bicarbonate reabsorption

c) type IV is due to decreased aldosterone secretion

d) type IV can be seen in patients on NSAID

e) RTA occurs in acute renal failure and in renal diseases

36. which of these statements is/are correct;

a) accuracy is the closeness of the test result to the true value

b) reliability is the test of how specific and sensitive a test is

c) sensitivity is the ability to detect that an analyte is absent when absent

d) specificity is the ability to detect an analyte when present

e) repeatability is a better measure of accuracy than reproducibility

37. Concerning errors in the scientific measurement,

a) errors can be avoidable(careless) or systematic

b) errors can be random or unavoidable

c) random errors are better controlled than systematic errors

d) systematic errors can be due to wrong labeling,wrong containers

e) the smaller the standard deviation the greater the precision.

38. Errors in the chemical pathology laboratory can be controlled by;

a) use of standards

b) use of accuracy control specimens

c) use of precision control specimen

d) use of control charts

e) use of sophisticated equipment

39. Concerning the Cerebrospinal fluid (CSF);

a) it is a transcellular fluid

b) the total volume is about 90-200ml in adults

c) can be obtained through venepuncture

d) high CSF pressure may be due to cerebral edema, SOL, meningitis, etc

e) low CSF pressure may be due to straining, holding breath, urination and prolonged cough.

40. Which of the following statements is/are true?

a) normal CSF is clear or colourless

b) Xanthchromic CSF may be due to altered hemoglobin, WBC, excess protein and jaundice

c) CSF protein can be assayed Biuret, Dye-binding, turbidimetry and immunological methods.

d) CSF transferring and prealbumin is 60-80% that in plasma

e) CSF protein continues to be higher until adulthood.

41. Which of the following is correct?

a) CSF/Plasma protein is preferred in determining blood brain barrier integrity

b) Oligoclonal bands on CSF electrophoresis suggests HIV, encephalitis, multiple sclerosis, epilepsy (after a fit)

c) CSF C-reactive protein level can be used to distinguish between viral and bacterial meningitis

d) CSF glucose is normally 80% that of plasma

e) CSF lactate differentiate between TB and viral meningitis and it is useful in monitoring unconscious patients.

42. The liver

a) Weighs 1300-1600g in adults

b) Held in place by coronary ligament, triangular ligament and falciform ligament

c) 60-80% blood supply is from the portal vein

d) Acinus zone I is richest in oxygen supply

e) Acinus zone III is most affected by anoxia and toxins

43. Functions of the liver include;

a) Glucose homeostasis and lipid metabolism

b) Synthesis of albumin, coagulation proteins and specific binding proteins

c) Storage of ascorbic acid and cobalamine

d) Synthesis of glycogen, triglycerides and lipids

e) Excretion of bile

44. Albumin

a) Synthesised in the liver and ref rangeis 32-55g/L

b) Half life is 20days with <5% turn over daily

c) Level indicates acute liver disease

d) Level is influenced by nutrition, plasma oncotic pressure and hormonal factors

e) Hypoalbuminaemias are usually hepatic in origin

45. Bilirubin

a) Exists as total, conjugated(lipid soluble) and unconjugated(water soluble) bilirubin

b) Conjugated form is seen in obstructive hepatic lesions

c) Conjugated hyperbilirubinaemia gives a positive bilirubin test in urine

d) Delta bilirubin is increased in obstructive jaundice

e) Increased total bilirubin without same for delta bilirubin is a good prognostic sign

46. Aminotransferases/Transaminases

a) ALT is more liver specific with highest conc in liver

b) AST has high conc in heart, RBC and muscle.

c) ALT is cytosolic while AST is both cytosolic and mitochondrial

d) ALT and AST levels correlate well with severity of liver damage.

e) ALT and AST are elevated in acute viral but not toxic liver injuries

47. Alkaline phospatase (ALP)

a) Derived from plasma membrane

b) Highest activity in liver, bone, intestine, placenta and kidney

c) Hepatic ALP rises due to biliary obstruction

d) GGT and 5'-nucleotidase determination is not necessary to localize ALP rise to the liver

e) Ectopic ALP include Regan(placenta-like) Kasahara and Nagao(intestine-like) isoenzyme

48. Clotting factors

a) liver is the site forsynthesis of I,II,V,VII,IX,X,XII,XIII

b) Prolonged PT and PTT time indicates normal liver

c) Before liver surgery and biopsy, PT and PTT should be done

d) Fibrinogen <g/L indicates severe liver damage

e) Calcium is necessary but not important

49. Components of electrophoresis include

a) Electricity source

b) Buffer

c) Template(starch,PAGE, agarose gel,etc)

d) Electrode and reading/detection mechanism

e) Stains...ponceau S, Sudan black, Oil red

50. Acute phase reactants (proteins)

a) Are plasma proteins

b) Readily available in the plasma

c) C-reactive protein is an example

d) Normal level is 60-80g/L

e) Are of no diagnostic significance

ISBN: 978-1-62265-970-8   ISBN-10: 1622659708 - Kay's Multiple Choice Questions in Chemical Pathology

Chapter 5   Answers

Block 1 Key

NO	A	B	C	D	E
1	F	F	T	T	T
2	T	T	T	T	F
3	T	T	F	F	T
4	F	T	T	F	F
5	T	T	F	T	T
6	F	F	T	F	T
7	F	F	T	T	T
8	T	F	T	F	F
9	F	T	F	F	T
10	T	T	T	F	T
11	T	T	F	T	T
12	F	T	T	T	T
13	T	F	T	F	F
14	F	F	T	F	T
15	F	T	T	T	F
16	F	T	T	T	T
17	F	T	F	T	T
18	T	F	F	T	F
19	T	F	T	F	T
20	F	T	T	T	T
21	F	F	F	F	F
22	T	T	F	T	F
23	F	F	F	T	F

24	F	F	F	T	T
25	F	T	T	F	T
26	T	T	T	F	F
27	T	F	F	T	F
28	F	F	T	F	T
29	T	T	T	T	T
30	T	T	T	F	T
31	F	T	T	F	T
32	T	T	T	F	F
33	T	F	T	T	F
34	T	T	T	F	F
35	F	T	T	T	T
36	T	F	T	F	T
37	T	F	F	F	T
38	T	T	F	F	F
39	F	T	T	F	T
40	T	F	F	T	F
41	T	F	F	F	F
42	F	F	F	F	T
43	T	T	T	T	F
44	F	F	F	T	F
45	T	T	T	T	T
46	T	T	F	T	F
47	T	F	T	F	T
48	T	T	T	T	T
49	T	T	T	T	T

50	F	F	F	F	F

Block 2 Key

	A	B	C	D	E
1	T	T	F	F	F
2	T	F	F	F	T
3	T	T	T	F	F
4	T	T	T	F	F
5	T	T	T	F	F
6	T	T	F	T	F
7	T	T	F	T	F
8	T	T	T	F	F
9	T	T	F	F	T
10	F	F	T	T	T
11	F	F	T	T	T
12	F	F	F	T	T
13	F	T	F	T	T
14	F	F	F	T	T
15	T	F	F	T	F
16	T	T	T	F	F
17	F	F	F	F	T
18	T	F	T	T	F
19	F	T	T	T	T
20	T	F	T	T	T
21	T	T	F	F	T
22	T	T	T	T	T
23	T	T	T	F	T
24	T	T	F	T	T
25	T	F	F	F	T
26	T	F	F	T	T
27	T	F	T	T	T
28	F	F	T	T	F
29	F	F	F	T	T
30	T	T	T	F	F
31	T	T	F	T	T
32	T	F	T	F	T
33	T	T	T	T	T
34	T	T	T	F	T
35	T	T	F	F	T
36	T	F	F	T	T
37	F	T	T	F	F
38	T	F	T	F	T
39	F	F	T	F	T
40	F	T	T	T	F
41	T	F	T	T	F

42	T	F	F	T	T
43	F	T	F	T	T
44	T	F	F	T	F
45	T	F	T	T	F
46	T	F	T	F	T
47	F	T	T	T	F
48	T	T	F	F	T
49	F	T	T	F	T
50	T	F	T	F	T

Block 3 Key

NO	A	B	C	D	E
1	T	F	F	T	F
2	T	T	T	T	T
3	T	T	T	T	T
4	F	F	F	T	T
5	T	F	T	F	F
6	T	T	T	T	T
7	F	T	T	T	T
8	T	T	F	F	T
9	F	F	T	F	F
10	T	T	T	T	F
11	T	F	T	F	T
12	T	F	T	T	F
13	F	F	F	T	F
14	F	F	F	T	T
15	F	T	T	F	T
16	T	T	T	F	F
17	T	F	F	T	F

18	F	F	T	F	T
19	T	T	T	T	T
20	T	T	T	F	T
21	F	T	T	F	T
22	F	F	F	T	T
23	T	F	T	T	F
24	T	T	T	F	F
25	F	T	T	T	T
26	T	F	T	F	T
27	T	F	F	F	T
28	T	T	F	F	F
29	F	T	T	F	T
30	T	F	F	T	F
31	T	F	F	F	T
32	F	F	F	F	T
33	T	T	T	T	F
34	T	T	T	F	T
35	T	T	T	T	T
36	T	T	F	T	F
37	T	F	T	F	T
38	T	T	T	T	T
39	T	T	T	T	T
40	T	T	T	T	T
41	F	F	F	F	F
42	F	F	F	T	T
43	T	T	F	F	T

44	T	F	F	T	F
45	T	T	T	T	T
46	T	T	T	T	T
47	T	F	T	F	F
48	T	F	F	T	F
49	T	F	F	F	F
50	F	F	F	T	F

Block 4 Key

	A	B	C	D	E
1	T	T	F	F	F
2	T	T	T	T	T
3	F	F	F	F	F
4	T	T	T	F	T
5	F	F	T	T	T
6	T	T	F	F	F
7	T	T	T	F	F
8	F	F	F	F	F
9	T	T	T	T	T
10	F	T	F	T	F
11	F	F	F	T	F
12	T	T	T	T	T
13	T	F	T	F	T
14	F	F	F	F	F
15	F	T	F	F	F

16	F	T	F	F	F
17	T	F	F	F	F
18	F	F	T	F	F
19	F	F	T	T	F
20	T	F	T	F	F
21	T	T	F	F	F
22	T	F	T	F	F
23	T	T	F	F	T
24	F	F	F	F	T
25	F	F	F	T	F

	A	B	C	D	E
26	T	F	T	T	F
27	T	F	F	F	F
28	F	T	T	T	F
29	T	F	F	T	F
30	T	T	T	T	F
31	T	F	F	T	F
32	T	T	F	F	T
33	T	T	F	F	T
34	T	T	T	F	T
35	T	T	T	T	F
36	T	T	F	F	F
37	T	T	F	T	T
38	T	T	T	T	F
39	T	T	F	T	F

ISBN: 978-1-62265-970-8   ISBN-10: 1622659708 - Kay's Multiple Choice Questions in Chemical Pathology

40	T	T	T	F	F
41	F	T	T	F	T
42	T	T	T	T	T
43	T	T	F	F	T
44	T	T	F	T	F
45	F	T	T	T	F
46	T	T	T	F	F
47	T	T	T	F	T
48	T	F	T	T	F
49	T	T	T	T	F
50	T	F	T	F	F

About the author: KAYODE (Kay) ADEBAYO

KAYODE (Kay) ADEBAYO

MB BCh(Ife Nig), MSc & MHRS(Lagos Nig), MCRRA(Chicago)
kjadebayo@yahoo.com, 7736552438, [08023071025]

Academic/Professional Training: Obafemi Awolowo University, Ile-Ife, Nigeria (Nov. 1986 – December, 1992)

Qualifications: Bachelor of Medicine, Bachelor of Surgery (MB ChB) December, 1992
Master of Science In Pharmacology (Msc)December 1999
Master of Humanitarian and Refugee Law (MHRS) 2007
Master of Clinical Research and Regulatory Administration.
(MCRRA 2014)

Professional License:
1.   Medical And Dental Council of Nigeria (MDCN) (Fm 21, 932)
          (Permanent Registration, January, 1994) MDCN/R/27,197
2.   MDCN Additional Qualification No: AQ5788 /2005
3.   HIPPA Certified by University of Illinois at Chicago. 2014.

Postgraduate Training:
1.   Fellow, National Postgraduate Medical College of Nigeria, (FMCPath) Faculty of Pathology, November 2004.
2.   University of Lagos, Nigeria, M.Sc. (Pharmacology) December, 1999.
3.   University of Lagos, Nigeria, M.H.R.S. (Humanitarian &Refugee Law)2007

Others:
1.   Yaba College of Technology, Diploma in Science Technology, 1986
2.   Certificate in Computer Appreciation, Words, Excel and Data Analysis.

Membership of Professional Bodies:
1.   Member, Nigerian Medical Association (NMA)
2.   Member, Pharmacologists Association Of Nigerian (PAN)
3.   Member, Association of Pathologists of Nigeria (ASOPON)
4.   Member, National Association for The Study of The Liver
5.   Member, Nigerian Association of Clinical Chemists (NACC) an Affiliate of IFCC, Milano. Italy.   (FM 093/04)
6.   Associate Member, Institute of Biomedical Science, London (AIBMS 00376049).

Voluntary/Extra Curricular Activities:
•   Nigerian Delegate to the International Federation of Medical Students Association (IFMSA) Conference Lagos 1988.
•   Nigeria Delegate to the Federation of African Medical Students Games (FAMSA) Ghana, 1991.
•   Member of FAMSA Standing Committee for Population Activities (SCOPA), 1989 – 1992.
•   University of Ife Medical School Representative to National Academic, Administrative and sporting activities, 1987-92.

Awards and Honours
•   Best Student in Chemical Pathology. OAU, Ife, 1990.
•   Adamawa State Honours Award (NYSC), 1993/94.
•   Fufore Local Government Chairman's Award, Adamawa State, (NYSC) 1993/94
•   AEROS Foundation Grant for Development of new Vaccine for Tuberculosis, 2004.
•   Dr Alex E.Boyo Prize For The Best Student in Part 1 Fellowship Exam 2005
•   Van Slyke Travel Grant to AACC Conference in Chicago, USA, 2006.

- FDR Scientific Grant for Research in Molecular Biology in USA, 2008.

Research and Academic Works:
- Homocysteine Level and Criminality: an established Correlation? (on-going)
- Homocysteine implication in delayed wound healing of typeII diabetes. (on-going)
- Glycated Haemoglobin and Glycaemic control: the Abakaliki experience(on-going)
- Keloids: A Review (Term Paper In Toxicology) Presented to the Department of Pharmacology, University of Lagos, June, 1999.
- Diuretics and Diuretics Resistance in Congestive Cardiac Failure: Postgraduate Term Paper as above, June, 1999.
- Human Chorionic Gonadotrophin: Term Paper As Above, May, 1999.
- Prevalence of Diabetes Mellitus and Hypertension in Fufore Local Government Area, Adamawa State, Nigeria, 1995.
- Reactive Depression In A Patient With Primary Infertility: A Case Study In Ile-Ife, Nigeria, 1992
- A Study of The Prevalence of Protein Energy Malnutrition among Children Age 0 – 10 Years In Ife Central Local Government Area Osun State, Nigeria, 1991. (In press)
- Problems and Prospects of Storage in Nigeria, 1986.

Publications:
1.	**Adebayo KJ,** Popoola JO: Burkitts' Lymphoma: A Five Year Review In Ile-Ife, Osun State. Nigeria. *Ifemed Journal* 1991; 9 (2) 17-19.
2.	Adebayo KJ, Odun FC, Uchenwoke TS. Nutritional survey of Children in Ife central local Government of Osun state. Nigeria. Ifemed Journal. 1992; 6(3) 8-13.
3.	**Adebayo KJ,** Oluwatowoju IO, Ajuluchukwu JNA: Plasma Homocysteine Levels In Normal And Pre-Eclamptic Pregnancies in Nigerian Women. *J. Clin. Sc.*2004;4,2(12) 18-20.
4.	**Adebayo KJ** : Thyroid Hormone Levels In Pregnant Nigerian Women. *Ann. of Med. Res.*2006;1(12) 46-48.
5. Osunkalu VO, **Adebayo KJ,** Akanmu AS: Level of Homocysteine among Primigravidae attending the Antenatal Clinic of The Lagos University Teaching Hospital. *Ann. of Med. Res.*2006;1(12) 30-34.
6.	**Adebayo KJ,** Ajuluchukwu JNA,Oluwatowoju IO: Plasma Homocysteine: Establishing Reference Values for Healthy Adult Nigerians. *Arch. of Clin. Res.*2006; 1 (12) 25-28.
7.	**Adebayo KJ**: Continuing Medical Education in Chemical Pathology. *Arch. of Clin. Res.*2006;1: (12) 70-73.
8.	**Adebayo KJ**, Madu EF, Adebayo-Kay VC: Serum total Homocysteine Concentration in children and adolescents in Jos, Nigeria. *J Trop.Pediatr* 2008; 8, 54(4):282-3
9.	**Adebayo KJ**, Madu EF, Adebayo-Kay VC: Ethnic Considerations in Plasma Total Homocysteine concentrations in Children and Adolescents in Jos, Nigeria. *Journal of Applied and Basic Sciences,* 2006;4: 48-52.
10.	**Adebayo KJ**, Alalade TO, Adebayo-Kay VC : Testosterone Levels among Healthy Pregnant Women in Lagos, Nigeria. *Journal of Applied and Basic Sciences,* 2006; 4: 58-61.
11. Samuel so, Ohaju-obodo jo, Salami tat, **Adebayo kj**, Abah SO.(2009) Swine Influenza Pandemic Threat: A review of the literature. *J. Appl. Bas. Sci.* 5(1&2):1-8.
12.	Ajuluchukwu JNA, Oluwatowoju IO, **Adebayo KJ**, Onakoya A. Hyperhomocysteinaemia in Africans with CVD. *World J Life Sci. and Medical Research 2011; 1 (6):126 -132.*
13.	**Adebayo KJ**, Adebayo-Kay VC. Anion gap in health and diseases: A review. *Journal of Applied and Basic Sciences,* 2013, 7;1: 1-4.
14.	**Adebayo KJ**. Aluminium and the chronic renal failure patient: a review. *Journal of Applied and Basic Sciences,* 2013, 7; 1: 5 – 8.
15.	**Adebayo KJ**, Adebayo-Kay VC. Homocysteine and serum cholesterol in pregnant Nigerians. *African journal of Med and Health Sciences*. Online September 2014.
16.	**Adebayo KJ**. Effects of forced migration on electrolyte profile of refugees. International *journal of behavioural science and health*. Online September, 30 2014.
4.

Accepted For Publication:
1. Nutritional assessment of refugees at a refugee camp (ACR)
2. Testosterone Levels Among Healthy Males In Lagos, Nigeria. (ACR)
In Press:
- Effect of forced Migration on Biochemical Profiles of Refugees (in press)

- Stability of Fertility Hormones after Assay
- Maternal Third Trimester Testosterone Level and Fetal Gender : An Established Correlation For Male Sex?(in press)
- Fractional Excretion of Uric Acid in Pregnancy: an Index for Preeclampsia? (in press)
- Investigating Male Infertility (in press)
- Homocysteine level and cognitive function in Children and adolescents in Jos, Nigeria (in press)

Abstracts
1.   Ajuluchukwu Jna, Oluwatowoju Io, **Adebayo Kj.** Homocysteine: A "New" Cardiovascular Risk Factor: Preliminary Data From 52 Healthy Adult Nigerians, Oct, 2004.
2.   **Adebayo Kj** Ajuluchukwu Jna, Oluwatowoju Io. Plasma Homocysteine Reference Values For Adult Nigerians. Dec. 2004
3.   **Adebayo Kj,** Madu EF. Serum Total Homocysteine Concentrations In Children And Adolescent In Jos,Nigeria.(Mema Conference, Lebanon 2005)
4.   **Adebayo Kj.** Plasma Homocysteine Levels In Normal And Preeclamptic Pregnancies In Nigerian Women (Dgkl/Oglmkc Conference, Germany 2004.

Academic and Scientific Conferences Attended
1.   Update Course In Laboratory Medicine – West African College Of Physicians, Faculty Of Laboratory Medicine. March 2-5, 1998
2.   Intensive Revision Course In Pathology, National Postgraduate Medical College Of Nigeria, Faculty Of Pathology, March 5-14, 1998.
3.   Update Course In Laboratory Medicine – West African College Of Physicians, Faculty Of Laboratory Medicine. March 1 – 2, 1999.
4.   Intensive Revision Course In Pathology, National Postgraduate Medical College Of Nigeria, Faculty Of Pathology. March 3 – 10, 1999.
5.   Intensive Revision Course In Pathology, National Postgraduate
6.   Workshop On Research Methodology In Medicine National Postgraduate Medical College Of Nigeria And West African Postgraduate Medical College July 31 – August 4, 2000
7.   Workshop on Reading And Writing Of Scientific Paper By Editor Of British Medical Journal; Lagos. May, 2000.
8.   Annual Scientific Conferences, Nigeria Medical Association, July, 2001
9.   Annual Scientific Conferences, Lagos University Teaching Hospital Association Of Resident Doctors, 1998 – 2001
10.   Annual General Meeting and Scientific Conferences, Association Of Pathologists Of Nigeria.
11.   Annual Scientific Conferences, National Association for The Study Of The Liver. 1998 – 2001
12.   Training Workshop On Liver Biopsy By National Association For The Study Of The Liver, February 28, 2001
13.   Update Course In Laboratory Medicine, 15 – 16 February 2002 Lagos University Teaching Hospital Lagos Nigeria. Organized By The West African College Of Physicians (Faculty Of Laboratory Medicine)
14.   Research Methodology Courses 4-8th March 2002 Lagos Nigeria Organised By The Faculty Of Public Health, National Postgraduate Medical College Of Nigeria.
15.   Health Resources Management and Administration Course. 18 – 28th March, 2002 Lagos Nigeria. Organised By The National Postgraduate Medical College Of Nigeria.
16.   Seminar On "Dynamics of Medical Business Management, Finance and Investment 31st May 2002 Lagos Airport Ikeja, Lagos, Nigeria. Organised by FB Asset Management Ltd and Nigeria Medical Association Lagos State Branch.
17.   Update Course In Laboratory Medicine 10-15 February, 2003 Lagos University Teaching Hospital Lagos Nigeria Organised By The West African College Of Physicians (Faculty Of Laboratory Medicine)
18.   Update Course In Pathology 10-22 March, 2003 Lagos University Teaching Hospital Lagos Nigeria Organised By The National Postgraduate Medical College (Faculty Of Pathology).
19.   Participant at The IFCC/Beckman-Coulter Protein Conference, 30 & 31 May, 2003, Barcelona, Spain.
20.   Participant At Euromedlab Confrence 2003, Palace Of Congresses, Barcelona, Spain, June 1-5, 2003
21.   Working Visitation To Johannes Guthenburg University For Collaborative Study On Development of New Vaccine for Tuberculosis. Mainz, Germany 1-14 August, 2004

22.   Workshop on Accident and Emergency Preparedness, NOH, Lagos March 2005.

23.   American Association of Clinical Chemists' Meeting/Conference, Chicago, USA, 22-27th July 2006.

24.   American Association of Clinical Chemists' Meeting/Conference, San Diego, USA, July 2007.

25.   American Association of Clinical Chemists' Meeting/Conference, Washington DC, USA, July 2008.

26.   Training and Laboratory Tutelage on Molecular Medicine Techniques. Merrillville, Indiana, USA. June/July 2008.

27.   American Association of Clinical Chemists' conferences in USA between 2009 and 2013.

28.   Drug development, clinical trials and Pharmacovigillance. FDR Scientific Inc, IN. USA. 2009-2013.

29.   Clinical Research and Regulatory Administration training. Northwestern University translational science unit. 2010-2013.